S0-EGP-013

SLUGS

by Ellen Lawrence

Consultant:
Dr. Ian Bedford
Head of Entomology
John Innes Centre
Norwich, United Kingdom

New York, New York

Credits
Cover, © Lisa S./Shutterstock; 4, © Peter Noyce GBRS/Alamy; 5, © Marko TanasijeviD/Alamy; 6, © Jaap Schelvis/Minden Pictures/FLPA; 7, © Dave Pressland/FLPA; 8, © Des Ong/FLPA; 9TL, © ImageBroker/FLPA; 9, © Lisa S./Shutterstock; 10T, © N-Sky/Shutterstock; 10B, © ajt/Shutterstock; 11, © Arterra Picture Library/Alamy; 12, © ImageBroker/FLPA; 13, © Bildagentur Zoonar GmBH/Shutterstock; 13R, © Eye of Science/Science Photo Library; 14T, © RT Images/Shutterstock; 14B, © Andrew Darrington/Alamy; 15, © Bill Coster/FLPA; 16, © Miriam Heppell/Stockimo/Alamy; 17, © Juniors Bildarchiv GmBH/Alamy; 18, © Kim Taylor/Nature Picture Library; 19, © Ingo Arndt/Nature Picture Library; 20T, © Jasper Rimpau/Dreamstime; 20B, © Gil Wizen; 21, © Michael Murphy; 22L, © Ruby Tuesday Books; 22R, © Lisa S./Shutterstock; 23TL, © Christian Camus/Shutterstock; 23TC, © Lisa S./Shutterstock; 23TR, © Pyshnyy Maxim Vjacheslavovich/Shutterstock; 23BL, © Marc Goldman/Shutterstock; 23BC, © Peter Noyce GBRS/Alamy; 23BR, © I Need Boss/Shutterstock.

Publisher: Kenn Goin
Senior Editor: Joyce Tavolacci
Creative Director: Spencer Brinker
Photo Researcher: Ruth Owen Books

Library of Congress Cataloging-in-Publication Data

Names: Lawrence, Ellen, 1967– author.
Title: Slime travelers : slugs / by Ellen Lawrence.
Description: New York, New York : Bearport Publishing, [2019] | Series: Slime-inators & other slippery tricksters | Includes bibliographical references and index.
Identifiers: LCCN 2018014553 (print) | LCCN 2018020337 (ebook) | ISBN 9781684027408 (Ebook) | ISBN 9781684026944 (library)
Subjects: LCSH: Slugs (Mollusks)—Juvenile literature.
Classification: LCC QL430.4 (ebook) | LCC QL430.4 .L39 2019 (print) | DDC 594/.3—dc23
LC record available at https://lccn.loc.gov/2018014553

For more information, write to Bearport Publishing Company, Inc., 45 West 21st Street, Suite 3B, New York, New York 10010. Printed in the United States of America.

10 9 8 7 6 5 4 3 2 1

Contents

A Slimy Trail 4
Super-Slimers! 6
Gooey Body 8
A Slug's World 10
What's for Dinner? 12
Slime for Safety 14
Slug Babies 16
Growing Up 18
A Rainbow of Slugs 20

Science Lab 22
Science Words 23
Index 24
Read More 24
Learn More Online 24
About the Author 24

A Slimy Trail

It's early morning in a garden.

Something has left a long, sticky trail on a walkway.

The slime-maker is a slug!

The little creature has spent the night looking for food.

As the sun comes up, the slug slides under a rock.

What is slime, and how do you think it is helpful to a slug?

a slug's slime trail

Slugs belong to an animal group called mollusks (MAHL-uhsks), which includes snails, clams, and octopuses. Mollusks have soft bodies and no bones. Some mollusks have shells to protect their bodies.

Super-Slimers!

A slug **secretes** slime from its whole body.

The slime is made of water and gluelike **substances**.

Some of the slime is very watery and helps keep a slug's body from drying out.

Other slime is thicker.

It helps the slug slide over the ground and even cling to walls upside-down!

Your body produces a substance that's very similar to slug slime. Can you guess what it is?
(The answer is on page 24.)

Inside a slug's body are slime-making substances. To make slime, a slug's skin soaks up water. When the water combines with the gluelike substances, lots of gooey slime forms.

Gooey Body

A slug's body has three main parts—a head, a mantle, and a tail.

If a slug is under attack, it can pull its head inside the mantle.

On the underside of its body is a strong, muscular foot.

The muscles in the foot tighten and loosen, so that the slug can move.

Then it slides along on its trail of slime.

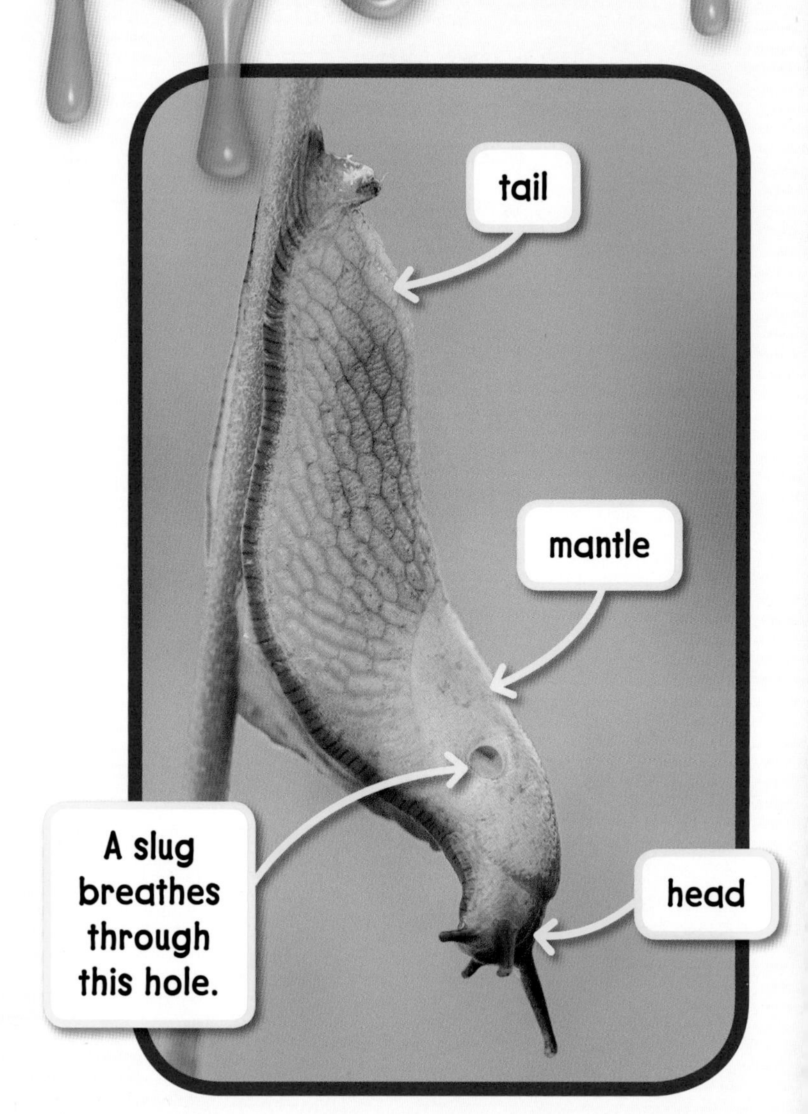

Where do you think slugs live and why?

A slug has two pairs of **tentacles**. The top pair contains the slug's eyes and is also used for smelling. The bottom pair is used for tasting and touching.

A Slug's World

Slugs live in fields, forests, and backyards.

They make their homes under fallen leaves, rocks, and rotting logs.

These places are cool and damp and help keep a slug's body moist.

Slugs also avoid the sun, only leaving their hiding places at night or on rainy days.

slugs on a rotting log

Many slugs actually live underground in soil.

A slug's slime trail does more than help it move. How else do you think a slug uses slime?

What's for Dinner?

When night falls, slugs leave their homes to look for food.

Some slugs eat leaves, flowers, mushrooms, and **lichen**.

Others feed on snails, worms, and even other slugs!

Many slugs eat dead, rotting plants and animals.

This helps clean up the places where they live.

As they feed, slugs leave a slime trail, which they use to find their way back home!

a slug eating a worm

A slug has a ribbonlike mouthpart called a radula (RA-joo-luh). It's covered with about 27,000 tiny, toothlike spikes. A slug shreds its food with its radula.

a close-up view of the toothlike spikes on a slug's radula

a slug eating a leaf

Slime for Safety

Many animals feed on juicy slugs.

Foxes, toads, frogs, and some birds and snakes will eat slugs.

A slug can produce extra slime to protect itself against **predators**.

The slime makes the slug slippery and hard to swallow.

The sticky goo makes some hungry predators think twice about snacking on a slug.

a toad hunting a slug

a bird catching a slug
Before eating a slug, some birds wipe their slimy meal on grass to remove as much of the goo as possible!

Slug Babies

When it's time to **mate**, slugs can follow slime trails to find other slugs.

All slugs are both male and female.

After a pair of slugs has mated, they both lay between 50 and 400 eggs.

Slugs lay their eggs in hidden, damp places, such as the bottom of a log.

This keeps the eggs from drying out and prevents predators from finding them.

a slug following a slime trail

A slug uses its tentacles to sniff out another slug's slime trail. The first slug can tell in which direction the other slug was traveling.

a slug with its eggs

Growing Up

A tiny baby slug hatches from each egg.

The young slug starts looking for food right away.

It takes about three months for a slug to grow to its adult size.

Then it's ready to lay eggs of its own.

A backyard may be home to thousands of slugs.

a tiny slug growing inside its egg

slug eggs
a baby slug that has just hatched

A Rainbow of Slugs

There are thousands of different kinds of slugs.

A leopard slug has dark spots like a leopard.

The leaf-veined slug looks like a leaf.

The giant pink slug can grow to 8 inches (20 cm) long.

Whatever they look like, all slugs are super-slimers!

How do you think the pattern on a leaf-veined slug helps it survive?
(The answer is on page 24.)

leopard slug

leaf-veined slug

Giant pink slugs live in only one place—on Mount Kaputar in Australia.
giant pink slug

Science Lab

Be a Slug Scientist!

Make a slug trap to find out what kinds of slugs live in your neighborhood.

1. Draw a small circle about 1 inch (2.5 cm) across on the lid of a plastic container. Then draw a star shape inside it. Ask an adult to help you cut along the lines of the star. Then push the spikes in.
2. Put the slug bait in the container and attach the lid.
3. Place the trap outdoors overnight. Slugs will climb into the container to reach the food, but they will not be able to escape.
4. In the morning, study any slugs you've trapped. Then release the slugs in a shady spot. Try not to touch them!

You will need:

- A small plastic container with a lid
- A pair of scissors
- Some slug bait, such as cat or dog food, or oatmeal mixed with milk
- A notebook, pencil, and ruler

Slug Investigation

- **How many slugs did you trap?**
- **How big were the slugs?**
- **What colors or patterns did the slugs have?**

Write all your observations in a notebook and then draw or photograph the slugs.

- **Go online and try to identify the slugs you've caught.**

Science Words

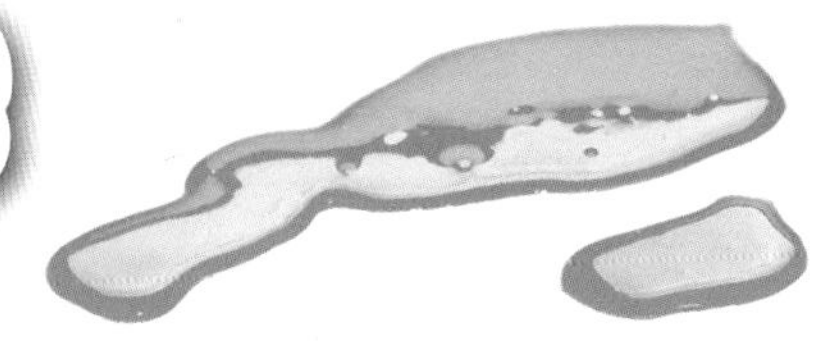

lichen (LYE-kuhn) small, plantlike living things that grow on rocks and trees

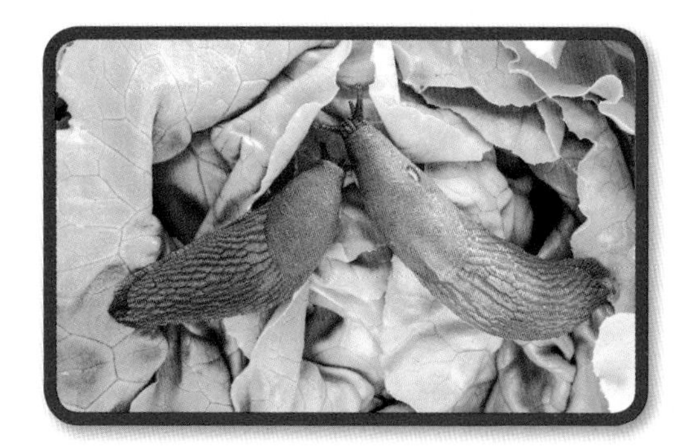

mate (MAYT) to come together to have young

predators (PRED-uh-turz) animals that hunt other animals for food

secretes (si-KREETS) releases, or squeezes out, a liquid or other substance from the body

substances (SUHB-stuhns-ez) particular kinds of liquids or other materials that often contain chemicals

tentacles (TEN-tuh-kuhlz) long body parts used by some animals for exploring their world

Index

baby slugs 18–19
eggs 16–17, 18–19
eyes 8–9
food 4, 12–13, 18
foot 8–9
mantle 8
mating 16
mollusks 5
predators 14–15, 16
radula 13
slime trails 4–5, 8, 11, 12, 16
tentacles 8–9, 16

Read More

Bishop, Celeste. *Slimy Slugs (Icky Animals! Small and Gross).* New York: Rosen (2016).

Borgert-Spaniol, Megan. *Slugs (Blastoff! Readers Level 1).* Minnetonka, MN: Bellwether (2016).

Riehecky, Janet. *Slime, Poop, and Other Wacky Animal Defenses (Blazers).* North Mankato, MN: Capstone (2012).

Learn More Online

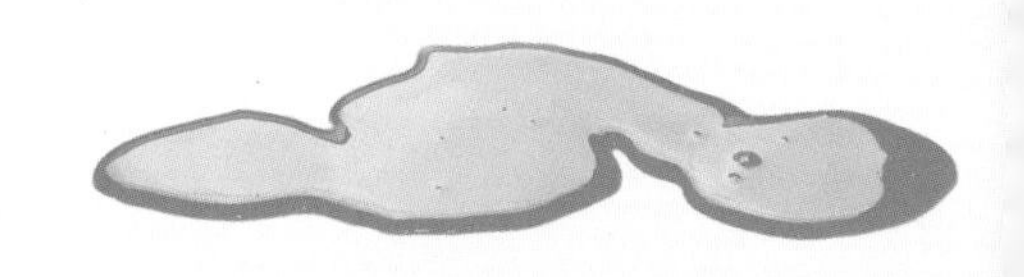

To learn more about slugs, visit
www.bearportpublishing.com/Slime-inators

About the Author

Ellen Lawrence lives in the United Kingdom. Her favorite books to write are those about nature and animals. In fact, the first book Ellen bought for herself when she was six years old was the story of a gorilla named Patty Cake that was born in New York's Central Park Zoo.

Answers

Page 6: Your nose produces a kind of helpful slime known as mucus. Sometimes, the air you breathe contains germs or tiny pieces of dirt. The mucus in your nose traps these harmful invaders and keeps them from getting into your throat and lungs.

Page 20: Scientists think that the pattern on a leaf-veined slug is a case of mimicry. Mimicry is when something looks like something else. Looking like a leaf, instead of something to eat, keeps the slug safe from predators, such as birds.